WORKING AT A ZOO

BY
BERTRAM T. KNIGHT

ℂ Children's Press
A Division of Grolier Publishing
New York London Hong Kong Sydney
Danbury, Connecticut

Created and Developed by
The Learning Source

Designed by Josh Simons, SimonSays Design!

Photographs by Tom & Therisa Stack

A CIP catalog record for this book is available
from The Library of Congress

ISBN 0-516-20751-2 (lib. bdg.)
      0-516-20377-0 (pbk.)

© 1998 by Children's Press®,
a division of Grolier Publishing Co., Inc.

Printed in the United States of America
1 2 3 4 5 6 7 8 9 10 R 07 06 05 04 03 02 01 00 99 98

What if your best friend was a shy giraffe or a bossy ostrich? The man with the giant tortoise wouldn't mind at all. That's because he is the **curator** of the Miami Metro Zoo.

The curator oversees all the animals at the zoo. But no one can do such a big job alone. It takes dozens of people to run a habitat zoo like this one.

Every zoo has **keepers,** and each keeper is responsible for certain animals. A good zookeeper becomes a trusted friend to the animals under his or her care—even a huge rhinoceros.

One of a zookeeper's
most important tasks is
to feed the animals. Some,
such as these kudus (*above*),
are fed leaves on long tree
branches. Lemurs (*right*) are
given their meals on a tray.

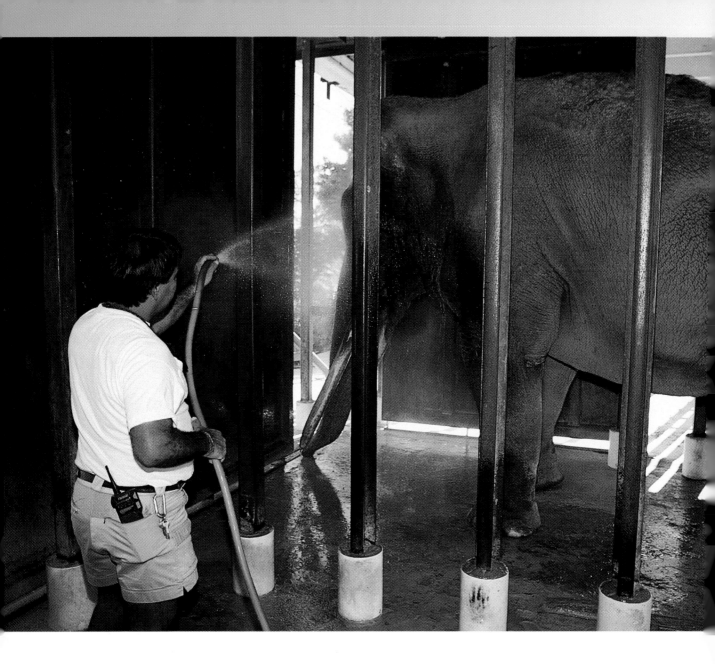

Zookeepers must also keep their animals clean. Can you imagine giving this fellow a bath day after day?

Welcoming newcomers is also part of a keeper's job. Many animals have traveled from far away and are scared and lonely. A caring keeper will soon make this white tiger feel right at home.

Just like people, animals need love. So keepers spend hours petting and playing with the animals in their charge. This koala loves being cuddled.

Other animals might not seem so cuddly at first. Still, this keeper spends lots of time talking to and nuzzling the camel he cares for.

A busy animal is a happy animal. So it is up to the keepers to provide plenty of exercise and play. This zookeeper is setting up a pumpkin hunt for his elephants. Another has built a special rope gym for the orangutans.

Keepers don't spend all their time with animals.
This senior keeper does lots of paperwork to make
sure that every animal receives the proper care.

Even the most well-tended animal can get sick or hurt. That is when a worried keeper contacts the zoo's **veterinarian,** or animal doctor.

Every zoo usually has at least one veterinarian, or "vet," on staff. Besides giving regular check-ups, vets handle illnesses and injuries. A picture from an X-ray machine will help this vet set the lemur's broken bone.

The Miami zoo also has labs where **veterinary technicians** study animal diseases and behavior. Information from these labs helps vets and keepers treat their animals.

Inside the zoo's offices, **inventory workers** keep records that tell where each animal comes from, its health, eating habits, and age. Anything that happens to an animal is noted on these records.

Food is big business at the zoo. In the wild, animals naturally eat what is good for them. But here, it is up to the **food staff** to provide the animals with the right diet.

The words "you are what you eat" are true for animals as well as humans. The staff carefully plans and then prepares each animal's meals— cleaning, chopping, and sometimes even cooking the food.

Dozens of office workers keep the zoo running smoothly. Some order the zoo's equipment and supplies. **Accountants** and **clerks** pay the bills for food, medicine, and staff salaries.

Another important part of zoo work is planning for the future. What new animals should be brought to the zoo? Does the reptile house need remodeling? Questions like these are handled by the **committee of planners.**

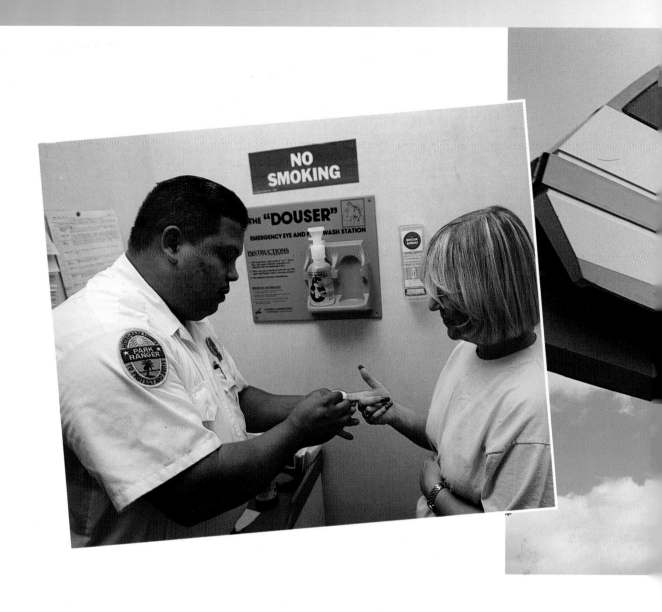

Security officers keep visitors and animals at safe and respectful distances from each other. Sometimes a security officer helps in other ways, too.

This monorail gives visitors a bird's-eye
view of animals in their habitats. The **monorail
operator** also guides the tour and describes
each exhibit along the way.

Every habitat and exhibit has signs telling visitors about the animals inside. Scientists called **zoologists** provide the information for these signs.

Clouded Leopard

Neofelis nebulosa

Clouded leopards are named for the large, irregularly shaped markings on their sides and back, which are thought to resemble clouds. Shy and elusive, their habits in the wild are not well known. An adult can weigh over 50 lbs., and over half of its 6 ft. length is its tail.

Their huge canine teeth are longer than other species of cats, relative to their size. In this regard they resemble the extinct sabre-tooth tiger. Highly arboreal, they are expert climbers and can pounce on deer from overhanging branches. They also feed on monkeys, squirrels, and birds.

Then **artists** create pictures to go with the information. With skillful hands, the artists draw animals, maps, and diagrams to help visitors find their way.

Zoos have **maintenance workers**, too. These
people do everything from keeping the grounds
neat to making sure that palm trees are trimmed
and healthy.

If a baby elephant is born or if a new exhibit is about to open, the **communications director** gives the news to newspapers, magazines, TV stations, and the public. He might also help put together a book like this one.

Who visits the zoo without a stop at the gift shop? The **shop's workers** don't just mind the store. They also pick out the best items to sell, order them, and make sure that the shop is a pleasant, interesting place.

It takes all these people—and more—to keep a place like the Miami Metro Zoo running. If you like to be around animals then maybe one day, you too, will be **Working Here**.

Taking a Closer Look

Page 3

In habitat zoos, animals seem to roam freely in large spaces. The habitats, though, are always surrounded by fences or ditches that keep the animals inside.

Page 5

Tortoises live on land. Unlike water turtles, which have flippers and webbed feet, tortoises have thick, stumpy hind legs. Some tortoises live to more than a hundred years old.

Page 4

Most people believe that giraffes are silent. In truth, giraffes actually are able to "speak" several different soft sounds.

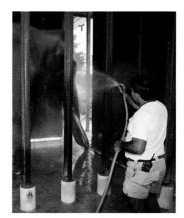

Page 8

This elephant does not live here. Along with other animals, it comes to this special stall only for a refreshing bath.

Page 4

The ostrich is the largest living bird. It cannot fly, but it is able to run at speeds of up to 40 miles (64 kilometers) per hour. Contrary to opinion, ostriches do not hide their heads in sand when frightened.

Page 9

Animals are usually flown to their new zoo homes in cages. Soon this white tiger will be released to roam comfortably in its new habitat.

Page 10

Although koalas may look like cuddly teddy bears, they are not bears at all. They are marsupials, similar to kangaroos and opposums.

Page 18

The animals' meals may be prepared by the food staff. But animal dieticians and vets usually plan what each animal will eat.

Page 14

The name rhinoceros comes from the Greek and means nose-horned. Depending on what kind it is, every rhinoceros has one or two horns growing from its long nose.

Page 24

There are many different jobs in the field of zoology. Some of them are managing wildlife preserves, heading expeditions, running labs or museums, and teaching.

Page 15

It takes most people four years of college and at least four more years of school beyond that to become a veterinarian.

Page 23

A monorail is a train that runs on only one rail. It can run either on the ground or overhead, like this one.

Index